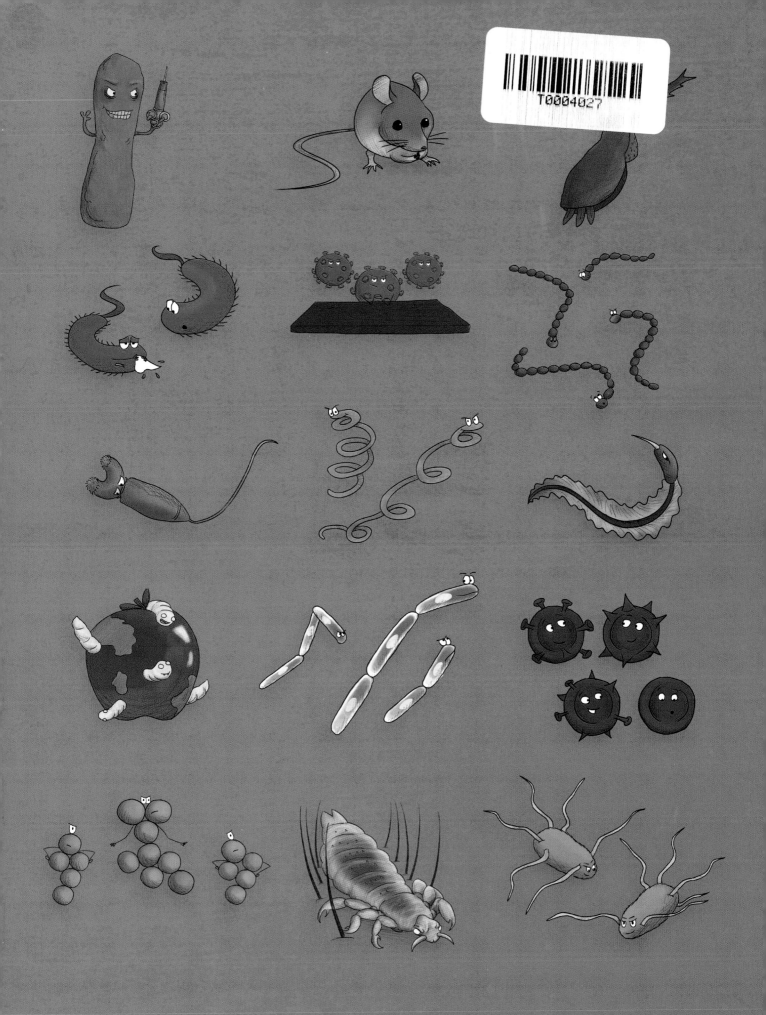

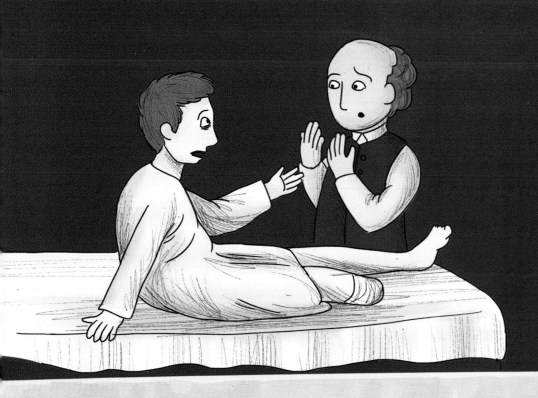

First published in 2022
by Hungry Tomato Ltd
F1, Old Bakery Studios, Blewetts Wharf,
Malpas Road, Truro, TR1 1JQ
United Kingdom

Beetle Books is an imprint of Hungry Tomato.

ISBN 978 1 913 440 50 3

Printed and bound in China

Discover more at:
www.mybeetlebooks.com
www.hungrytomato.com

CONTENTS

Grisly Medicine   8

Words in bold can be found in the glossary.

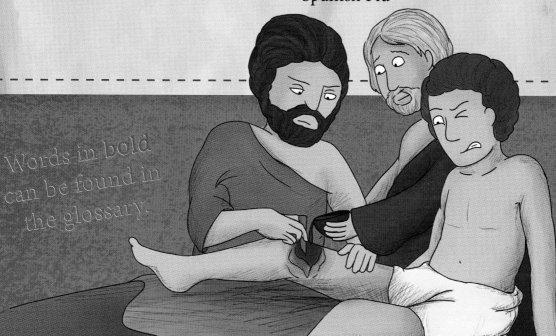

GRISLY MEDICINE

By John Farndon

Illustrated by Venitia Dean

GRISLY MEDICINE

It's not nice when you get sick. While most doctors try to make you feel better, in the past their methods were sometimes weird, yucky or even dangerous. Explore the microscopic monsters that make us sick, some of the great ideas that have made us better, and some not so great ideas...

Is there a doctor in the house?

There are between 10 and 15 million doctors in the world. People in Spain have more doctors each than anywhere else in the world. The World Health Organization thinks that we really need over 4 million more doctors.

Catching germs

Germs are found all over the world, in all kinds of places. Some reach you through the air via sneezes, coughs or someone's breath. You can also become infected if you touch somewhere contaminated with germs, then touch your nose and breathe in. Other germs can be swallowed when you eat food that has gone bad.

Germ killers

Some medicines are designed to help people get better by fighting infections. Antibiotics, for instance, are medicines that kill or attack the germs that make people ill. **Antibiotics** only work against **bacteria**. To fight **viruses**, you need special antiviral drugs.

The worst outbreaks

Doctors today call the worst outbreaks of disease '**pandemics**'. These spread far and wide, killing millions. Throughout history there have been many pandemics, including the recent Coronavirus outbreak. One of the worst pandemics was the Black Death of 1346-53, which was carried and passed on by fleas and killed 75-200 million people worldwide.

God's vengeance

One of the frightening things about diseases in the past is that people had no idea what caused them. Now we know diseases are spread by germs – tiny bacteria and viruses – so we can look for ways to fight them. But in the past there was no explanation. Many people simply believed diseases were caused by angry gods.

SETTLING FOR GERMS

In the early days of mankind, when people roamed around hunting and gathering, infectious diseases were probably rare. People did not live close enough together for germs to spread, or stay long enough near water sources to pollute them.

Disease on the farm

The development of farming some 10,000 years ago saved mankind from starvation and provided food for the first towns and cities. But we now know that germs multiplied as people and farm animals began living close together, sharing their germs, and water became polluted.

Slow-moving water provided a breeding ground for parasites.

Cattle gave us tuberculosis and smallpox.

Dogs gave us measles.

Chickens may have given us flu.

The first known polio victim?

On an Egyptian monument from the 13th century BCE, there is a picture of a man called Roma. He seems to have a withered leg, which could have been caused by polio. If so, he is the first known victim of this terrible disease, which may have developed in water dirtied by the dung of farm animals.

As farming became more intense, manure-polluted water encouraged diseases such as polio, cholera, typhoid and hepatitis. Slow-moving irrigation water provided ideal conditions for parasites, such as those causing the diseases bilharzia and malaria.

Horses gave us the common cold.

Polio may have developed in water dirtied by farm animal dung.

Pigs may have given us flu.

11

EASTERN IDEAS

You might think people had no idea about germs in the past because they are just too small to see. But in India, more than 2,500 years ago, those who followed the Jainist religion were taught that tiny life forms called 'nigoda' existed all around them.

Saving germs

Just as people nowadays often wear surgical masks in order not to breathe in germs, so did the Jains, thousands of years ago. However, their aim was not to avoid disease. They believed they must not harm any living thing, and wanted to avoid killing the tiny nigoda by accidentally swallowing them!

Ancient medicine

Ayurveda is an ancient medical system, developed in India more than 3,000 years ago, that uses complex mixes of herbs to treat people. Scientists are not sure if it works, but many people still use it. More than 2,000 years ago, Ayurvedists believed that **microbes** could cause diseases, such as leprosy and meningitis.

Prevention better than cure

The most famous old book of Ayurvedic medicine was the *Charaka Samhita,* written about 2,200 years ago by Charaka. He argued that prevention is better than cure. That's why he suggested different diets to keep you healthy in different places and at different times of year.

The plague in Granada

In 1350, the bubonic plague reached Granada in Spain, then part of the Islamic world. Physician Ibn Khatima suggested it was spread by 'minute bodies', which sound rather like germs. Another physician, al-Khatib – shown above in the orange headscarf, with the Vizier (ruler) in Granada's Alhambra Palace – explained how such organisms spread the plague by contact between people.

INVISIBLE ANIMALS

Around 1590, Dutch spectacle maker Zacharias Janssen put some lenses together – and invented the microscope. When people began to look through it, they were blown away by what they saw: a whole new unknown world of tiny organisms, too small to see normally.

Hooked in

In 1665, the English physicist Robert Hooke (1635–1703) published *Micrographia*, a book full of drawings of the amazing things he viewed down a microscope – things people had never seen before.

In slices of cork, he could see a honeycomb network of boxes. He called the boxes 'cells', and we now know all living things are built up from tiny cells.

Tiny monsters

Dutch scientist Anton van Leeuwenhoek (1632–1723) made a microscope with just a single lens. It was simple but brilliant, and he could see things 200 times larger than life! He discovered that clear water is not clear at all, but teeming with tiny creatures. In fact, there are tiny creatures almost everywhere!

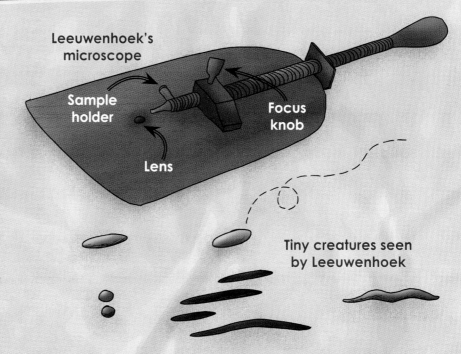

Leeuwenhoek's microscope

Sample holder

Focus knob

Lens

Tiny creatures seen by Leeuwenhoek

Germs spotted

Leeuwenhoek was astonished by the range of creatures he saw (above). Many looked like monsters. It was Leeuwenhoek who saw germs for the first time, when he looked through his microscope at the bacteria wriggling in plaque taken from his wife's and daughter's teeth.

The bacteria four

After Leeuwenhoek's great discovery, no one paid much attention to bacteria for the next 200 years, until German biologist Ferdinand Cohn began to study them closely in the 1870s. Cohn decided that, though there are many kinds of bacteria, they can be divided into four groups, depending on their shape: spheres, rods, threads and spirals.

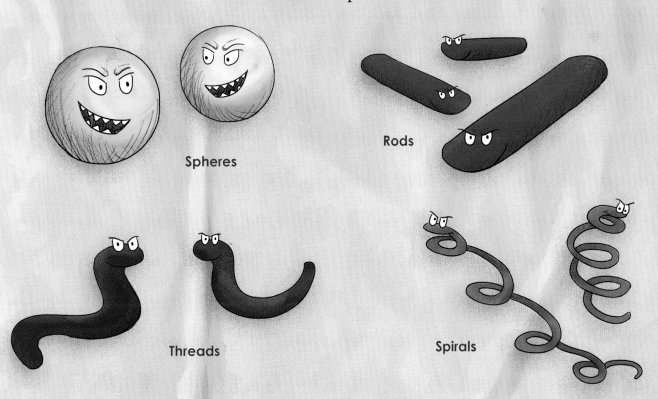

Spheres

Rods

Threads

Spirals

DIRTY DRAINS

In the 1800s, London and other European cities grew rapidly – and so did deaths from the killer disease cholera. Poor people were especially hard hit by cholera.

Feeling blue

Cholera is a horrible disease that seems to have originated in India on the banks of the River Ganges. Victims may suffer from diarrhoea so severe that their bodies are drained of water. Their eyes go hollow, their skin shrivels and they turn blue.

Where's that smell?

Experts in the past were convinced that cholera was spread by damp, smelly mist, known as **miasma,** found near ditches and swamps. A cartoon of the time pokes fun at health officials trying to sniff out just where the worst stinks were coming from.

The Soho pump

English doctor, John Snow didn't think the miasma theory was right. When cholera hit London's Soho district in 1854, he found that all the victims had drunk water from just one pump in Broadwick Street. Human poo had contaminated the water supply for the pump. However, it was a while before people understood that cholera is caused by germs in dirty water.

The fast way to travel

The opening of Egypt's Suez Canal in 1869 cut the sea journey from India to Europe dramatically – and made it much easier for cholera **epidemics** to spread. London escaped the worst because by then it had built good sewers, but many other European cities suffered badly.

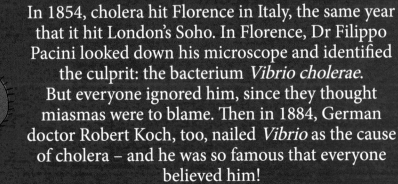

Vicious Vibrion

In 1854, cholera hit Florence in Italy, the same year that it hit London's Soho. In Florence, Dr Filippo Pacini looked down his microscope and identified the culprit: the bacterium *Vibrio cholerae*. But everyone ignored him, since they thought miasmas were to blame. Then in 1884, German doctor Robert Koch, too, nailed *Vibrio* as the cause of cholera – and he was so famous that everyone believed him!

GUILTY GERMS

In the 1840s and '50s, Semmelweis and Snow showed how clean hands and good drains could cut the spread of disease. So how could bad air be to blame? Then in the 1860s, the French microbiologist Louis Pasteur began to demonstrate that the real culprits are germs – also known as microbes.

Life from nothing

When people saw maggots crawling from rotten apples or dead meat, it didn't occur to them that the maggots had hatched from eggs. They thought they simply appeared whenever food rotted. This idea of life appearing from nothing is known as 'spontaneous generation'.

Life from the air

In 1859, Pasteur (1822–95) demonstrated with a simple experiment that microbes won't appear magically by spontaneous generation when food decays. Instead, they fall from the air on dust.

1. Pasteur boiled meat broth in a flask with a bent 'swan' neck to kill off any microbes.

2. The swan neck stopped dust falling in, so the broth stayed clear.

3. He then broke off the neck to allow dust to fall in. The broth quickly went cloudy, showing microbes were multiplying.

Germs on worms

In 1867, Pasteur showed that germs can cause disease. He had been studying a disease that was killing silkworms, the caterpillars that make silk. When Pasteur examined the diseased worms through his microscope, he found they were infected by not just one kind of microbe, but two.

Chicken test

In 1880, Pasteur discovered how to protect people against germs with experiments on chickens. He grew chicken cholera germs in a dish, then starved them of nutrition so they became weak. When he injected chickens with these old germs, the chickens became immune to the disease. Vaccinating people with weakened germs is now one of the main ways of preventing diseases.

Sheep shot

In 1881, Pasteur also found a way to weaken the germs that cause anthrax in sheep. Using these weak germs, he vaccinated sheep so that they became immune to anthrax.

THE OLDEST KILLER

Malaria is a really terrible disease that affects mostly tropical regions. It kills nearly half a million people every year and makes over 200 million ill. People become infected when a **mosquito** bite injects a tiny microbe into their blood.

Tiny assassin

Malaria is caused by the microbe called **plasmodium**. But it is spread by female Anopheles mosquitoes. When this mosquito bites someone infected with the disease to feed on their blood, it picks up the microbe. When it bites someone else, it passes on the microbe, infecting them with malaria.

An ancient disease

Malaria is the oldest known disease. The plasmodium microbes that cause it have been found in mosquitoes from as long ago as 30 million years. These mosquitoes were trapped in the resin that oozes from some trees, then perfectly preserved as the resin hardened and turned to **amber**.

Fever tree

Malaria was spread to the Americas by European settlers in the 1500s. But the American Indians learned to treat it using the powdered bark of the cinchona tree, which soon became known as the 'fever tree'. A drug called **'quinine'**, which is made from cinchona bark, is still an effective way of treating malaria.

Bad air

The word 'malaria' comes from the Latin for 'bad air', and for a long time people thought it was caused by the damp, smelly air given off by swamps. They weren't so far wrong, because these swamps are the perfect breeding ground for the mosquitoes that pass on the disease.

Smelly breath

Malaria became a killer disease as soon as people settled down to farm 10,000 years ago. The people who built the Egyptian pyramids stuffed themselves with garlic to ward off the disease. So they must have had very smelly breath! But scientists now think garlic really does help fight malaria.

29

DEADLY BLISTERS

In 541–42 CE, the city of Constantinople (now Istanbul) was utterly ravaged by an outbreak of a terrible disease called the Justinian plague, named after the city's ruler, Justinian. Up to 10,000 people died each day, and the streets were piled high with bodies.

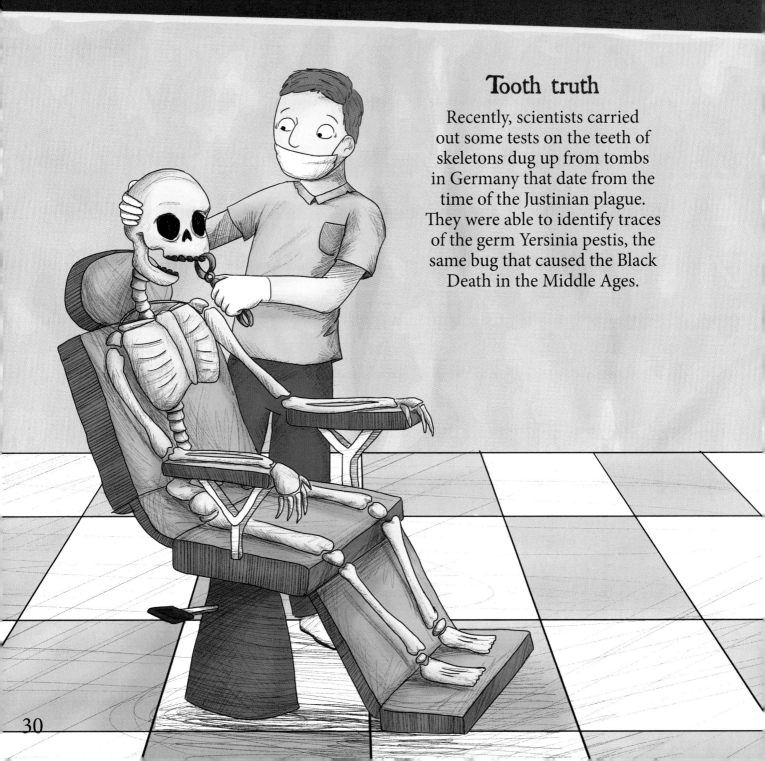

Tooth truth

Recently, scientists carried out some tests on the teeth of skeletons dug up from tombs in Germany that date from the time of the Justinian plague. They were able to identify traces of the germ Yersinia pestis, the same bug that caused the Black Death in the Middle Ages.

You look so well, ladies (not!)

If someone caught the plague, they would feel like they had the worst flu ever. Then, parts of their body would turn black and their skin would erupt with terrible pus-filled swellings called '**buboes**' – or worse still, their lungs would dissolve from the inside. Within a week they'd be dead...

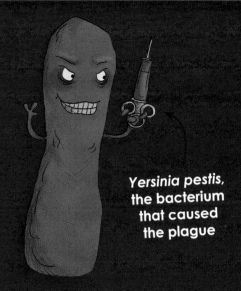

Yersinia pestis, the bacterium that caused the plague

That pesky *pestis*

Rats may have carried the disease to Constantinople, but the culprit was really a tiny bacterium called Yersinia pestis. Yersinia may be tiny but it was one of the deadliest killers in history. It also brought the Black Death in the Middle Ages, and the plague that killed millions in Asia in the later 1800s.

The march of death

The germs were carried to Constantinople by rats that stowed away on ships carrying grain from Egypt. From Constantinople, the plague spread rapidly, engulfing most of North Africa, the Middle East and Europe. Altogether, it is thought to have killed 25 million people in less than two years.

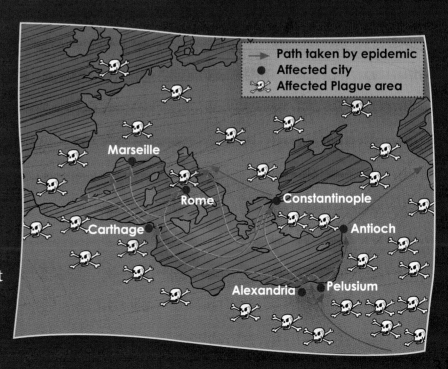

Path taken by epidemic
Affected city
Affected Plague area

Marseille
Rome
Carthage
Constantinople
Antioch
Alexandria
Pelusium

31

THE GREAT PLAGUE

The Black Death was the worst outbreak of disease in history. But for more than three centuries after, Europe was subjected to repeated plagues. The last massive outbreak was the Great Plague, which struck London in 1665.

Doctor Beak

It was a brave doctor who dared go near plague victims to tend to them. Because the disease was thought to be spread by bad air, some doctors dressed in a weird costume with a face mask and a long beak filled with herbs and flowers. They thought these might keep away the bad air.

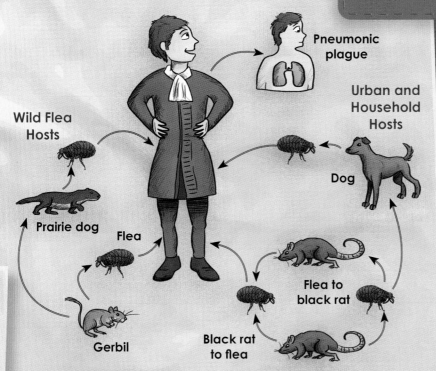

Pneumonic plague

Urban and Household Hosts

Wild Flea Hosts

Dog

Prairie dog

Flea

Flea to black rat

Gerbil

Black rat to flea

Plague pathway

Bubonic plague is caused by a bacterium that infects humans when they are bitten by a flea. These fleas live on rats. They can live on cats and dogs, too. Bubonic plague can then develop into pneumonic plague, which dissolves the lungs horribly and can be spread through the air in coughs and sneezes.

Little monsters

The Yersinia pestis bacterium that caused the bubonic plague was carried by fleas. But pneumonic plague could be spread from one human to another through the air. The Black Death and Great Plague were probably a mix of both kinds.

Stinky streets

In the 17th century, there were no proper drains in cities like London, and people just chucked the contents of their toilet out into ditches in the narrow streets – often barely missing people walking past! In these filthy conditions, infections of all kinds could spread easily.

THE WHITE PLAGUE

Bubonic plague is now, thankfully, mostly a terror of the past. But the horrible lung disease tuberculosis, or TB, once known as the 'white plague', is still with us, and millions of people around the world are affected by it.

Trust me, I'm an emperor...

The mythical Chinese Yellow Emperor, Huangdi, fancied himself a doctor. A medical book called the Huangdi Neijing, said to be based on his ideas, contains the earliest known description of TB. But it was written over 2,000 years after Huangdi's time – so someone must have had a very good memory!

Touch me, touch me!

TB doesn't just affect the lungs. It can cause swellings on the neck called 'scrofula'. In the Middle Ages, people with scrofula would queue up to see the king, because it was thought that being touched by the king would cure them. So the disease became known as the King's Evil.

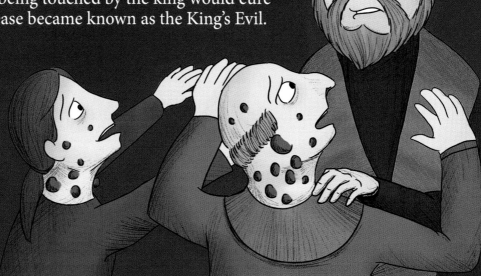

Symptoms of *Tuberculosis*

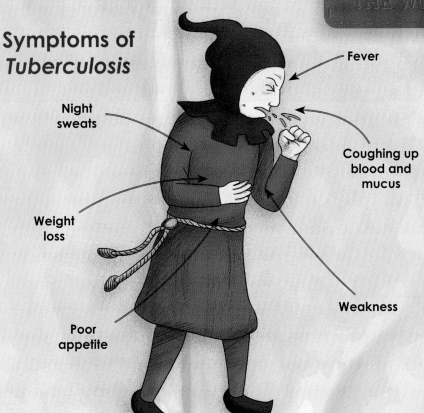

Fever

Night sweats

Coughing up blood and mucus

Weight loss

Weakness

Poor appetite

Coughs and fever

TB is a horrible disease that kills people slowly if they are not treated. It makes people cough terribly, often spitting up blood. It makes them sweat badly at night, and it makes them lose a lot of weight and become weak. This is why it came to be called **'consumption'**, because it seemed to consume the victim's body.

Vampire killers

People once thought TB victims had pale skin because vampires were sucking their blood and draining their lives away. That's why, when so many young girls died of consumption in the 1800s, writers wrote horror stories about vampire attacks.

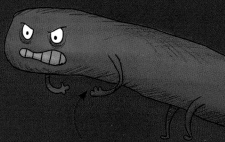

The *Mycobacterium tuberculosis* (MTB) bacterium, which causes TB

Romantic death?

In the 1800s, consumption claimed the lives of many young poets, such as John Keats, and many young girls. Their pale skin, for some, had 'a terrible beauty'. And so the disease came to seem almost romantic. But the victims suffered terribly, and their deaths caused their loved ones great heartache.

THE POX

Thanks to **vaccination**, the disease smallpox is dead. But for a long time in the past it was the world's worst killer, and even those who survived usually ended up with faces disfigured by the skin rash it caused.

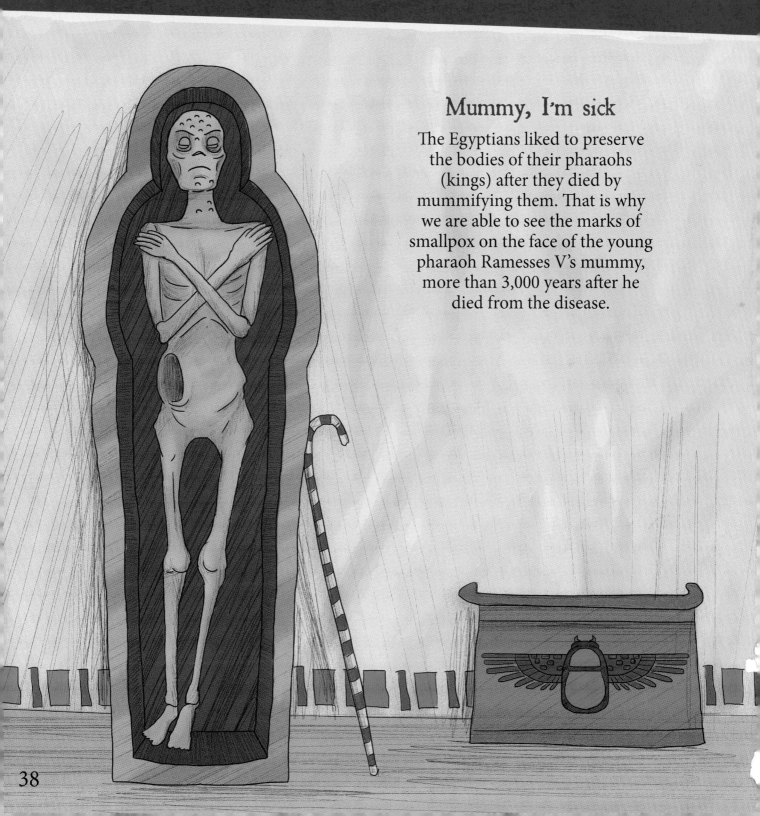

Mummy, I'm sick

The Egyptians liked to preserve the bodies of their pharaohs (kings) after they died by mummifying them. That is why we are able to see the marks of smallpox on the face of the young pharaoh Ramesses V's mummy, more than 3,000 years after he died from the disease.

White queen

In October 1562, the young Elizabeth I of England caught smallpox. For a week, her life hung in the balance. She pulled through, but her face was scarred by the disease and her hair fell out. For the rest of her life, she painted her face in thick white lead paint and egg whites, and wore wigs.

Nasty virus

Smallpox is caused by a virus called Variola. It jumped from rodents to humans 16,000 years ago, and learned how to invade body cells, causing smallpox. Thanks to vaccination, there are now just a few ariola viruses left – safely locked up!

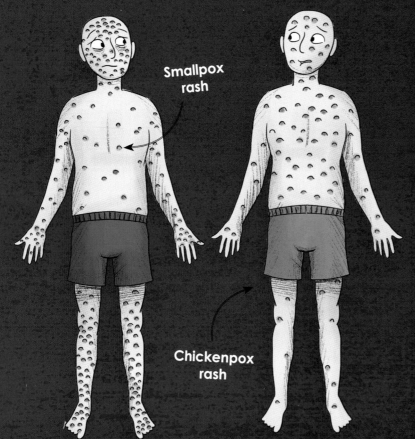

Smallpox rash

Chickenpox rash

Signs of smallpox

For two weeks after a person catches the smallpox virus, nothing happens. Then they start to feel like they have flu, but seem to get better. Suddenly red spots appear on their face and forearms, before spreading over their whole body and they become very ill. Smallpox can look like mild chickenpox, but the marks are much more dense.

39

NAPOLEON'S NIGHTMARE

Like smallpox, typhus has been almost eradicated, thanks to vaccination. But, like smallpox, typhus was once a deadly disease that brought a great deal of suffering and death. At first, victims suffered from an illness like flu, then developed a terrible rash over their whole body.

Beaten by a microbe

The French emperor Napoleon conquered most of Europe, but after reaching Moscow in 1812 and finding it abandoned, he decided to retreat. On the way back, through a bitter winter, more of his army was killed by typhus than by all the armies of Russia.

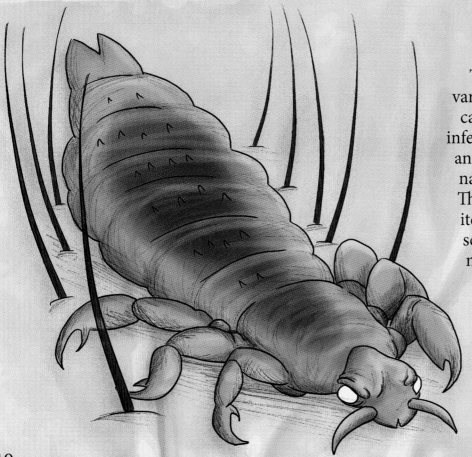

Nasty bugs

Typhus is caused by various kinds of bacteria called Rickettsia. They infect humans either from animal droppings or via nasty little lice or fleas. These can make people itch so much that they scratch their skin and make an opening for the bacteria.

Delousing

In the aftermath of World War II, typhus could easily have spread rapidly among closely packed soldiers and the hordes of refugees. But millions of lives were saved when people were sprayed with the newly discovered chemical, DDT, which killed the lice carrying the disease.

Sentenced to sickness

In the 1500s, many crime suspects died in prison, of typhus or 'jail fever', before they could be tried. In 1586, 38 fishermen accused of stealing fish were carried into court in Exeter, England, half dead of typhus – and spread the disease to all the court officials!

Irish fever

The poor in Ireland suffered especially from typhus at the time of the Irish famine in the 1840s. Many starved, and just as many were forced to leave their homes for other countries, when a disease killed off the potatoes that they relied on for food. From Ireland, typhus spread to England and became known as Irish fever.

41

CHOLERA

Cholera is one of the nastiest diseases there is, giving people diarrhoea and making them horribly sick. People become infected by drinking dirty water in which the bacterium Vibrio cholerae thrives.

New sewers

Sewage was often collected in tanks called cesspits or dumped in the river, making London super smelly. After Dr John Snow's discovery of the dangers of sewage contamination (right), London embarked on a massive task of drain building, to wash sewage clean away. Few people got cholera in the city ever again.

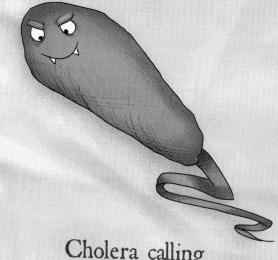

Cholera calling

The Vibrio cholerae bacterium that causes cholera thrives in water and food that has been contaminated by human poo. Humans can catch cholera if they eat sea creatures that swim in sewage-contaminated water. When people eat or drink anything containing cholera germs, they multiply in the gut and cause serious illness.

Snow's discovery

No one knew what caused cholera until an outbreak of the disease in London's Soho in 1854 prompted Dr John Snow to investigate. Snow found that victims had all drunk water from a pump in Broadwick Street. It turned out that the pump water was being contaminated by leaking sewage.

Flying the yellow flag

Cholera became widespread in the 19th century when towns grew, yet people had poor access to fresh water, and poor sewage systems. If anyone on board a ship got cholera, the ship had to fly a yellow-and-black flag to warn other ships. No one from the ship would be allowed ashore for a month.

YELLOW FEVER

Yellow fever is a dangerous tropical disease that makes people very ill with fever and vomiting. Sometimes it makes the skin turn yellow as the germs damage the cells of the liver.

Yellow for danger

In most cases, yellow fever only makes people sweaty and sick for a few days. But for one in six people, the virus then attacks the liver, making the skin turn yellow and causing a nasty abdominal pain. Then the mouth and eyes bleed, and so does the gut, making the sufferer vomit black blood.

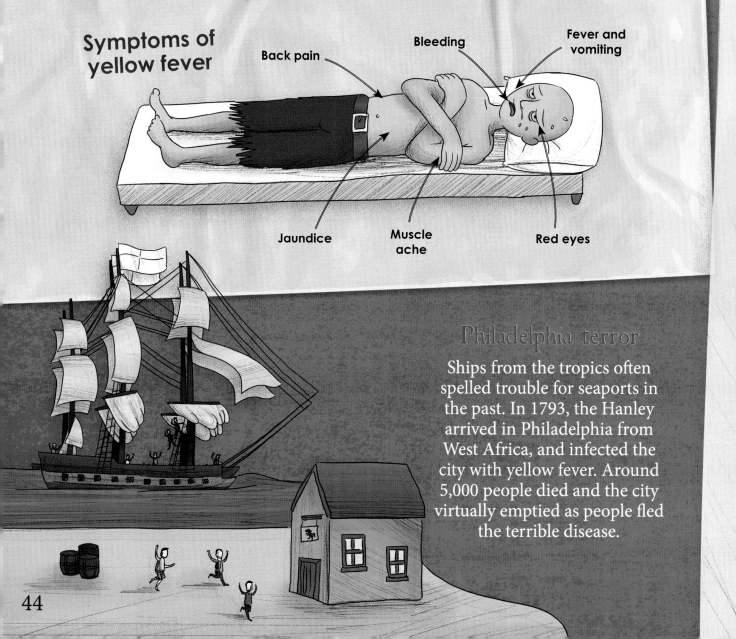

Symptoms of yellow fever

Back pain

Bleeding

Fever and vomiting

Jaundice

Muscle ache

Red eyes

Philadelphia terror

Ships from the tropics often spelled trouble for seaports in the past. In 1793, the Hanley arrived in Philadelphia from West Africa, and infected the city with yellow fever. Around 5,000 people died and the city virtually emptied as people fled the terrible disease.

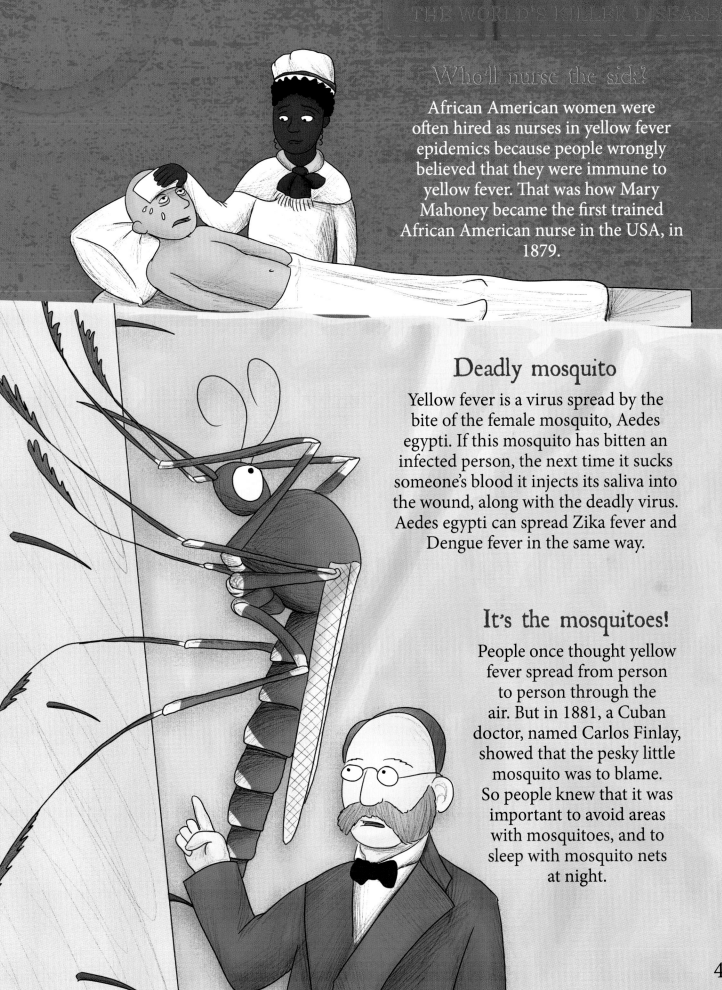

Who'll nurse the sick?

African American women were often hired as nurses in yellow fever epidemics because people wrongly believed that they were immune to yellow fever. That was how Mary Mahoney became the first trained African American nurse in the USA, in 1879.

Deadly mosquito

Yellow fever is a virus spread by the bite of the female mosquito, Aedes egypti. If this mosquito has bitten an infected person, the next time it sucks someone's blood it injects its saliva into the wound, along with the deadly virus. Aedes egypti can spread Zika fever and Dengue fever in the same way.

It's the mosquitoes!

People once thought yellow fever spread from person to person through the air. But in 1881, a Cuban doctor, named Carlos Finlay, showed that the pesky little mosquito was to blame. So people knew that it was important to avoid areas with mosquitoes, and to sleep with mosquito nets at night.

45

SPANISH FLU

Most of the time, flu is just a minor winter illness. But some forms are deadly killers. An outbreak of one particular kind, called Spanish flu, killed between 50 and 100 million people in 1918–19, a truly staggering number.

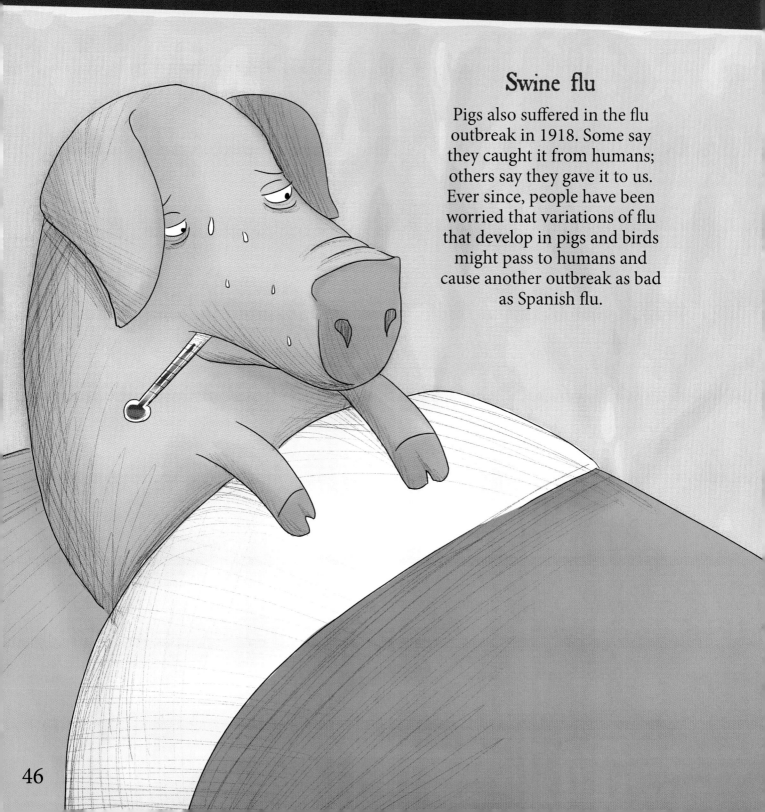

Swine flu

Pigs also suffered in the flu outbreak in 1918. Some say they caught it from humans; others say they gave it to us. Ever since, people have been worried that variations of flu that develop in pigs and birds might pass to humans and cause another outbreak as bad as Spanish flu.

Flu in the trenches

In World War I, millions of soldiers were crowded together in muddy trenches in France. Pigs were crammed up near them, too, to provide food. It's thought the flu virus developed its deadly form as it was passed back and forth between pigs and soldiers.

Sprayed in snot

Flu is very contagious, or easily spread. Someone with flu only has to sneeze for tiny droplets full of the virus to spray into the air and for someone else to breathe them in. During serious flu outbreaks, people may wear masks to avoid breathing in germs, but they are not always effective.

Camp Funston

Camp Funston might sound like a nice place to go on holiday. But in March 1918 it was anything but. It was a training camp for young soldiers in Kansas, USA, and is thought to be where the terrible Spanish flu epidemic began.

47

BODY BUTCHERS – THE ANATOMISTS

From the 1500s on, people thought it might help to know where things are actually positioned in our bodies (our anatomy) and how they work (our physiology). But finding out could be a rather nasty business!

Burke and Hare

In the early 1800s, criminals often dug up bodies from graves to sell to medical schools. But in Edinburgh in 1828, William Burke and William Hare found a way to speed up the process. They didn't wait for people to die; they just killed them and sold their bodies to Dr Robert Knox for his famous anatomy lectures.

Cut up

In the 1530s, Italian doctor Andreas Vesalius discovered that the only way to learn about human anatomy was to cut bodies open. He did this to make detailed, accurate drawings of what is inside. Cutting up bodies is called 'dissection'.
In 1543, he dissected the body of a criminal in front of a huge audience.

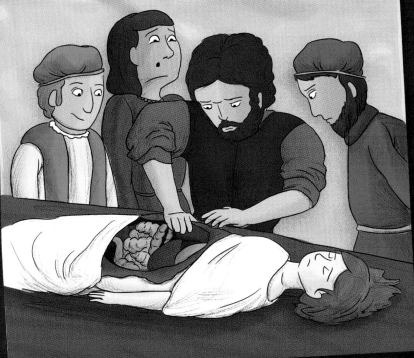

I said half a teaspoon...

Swiss German doctor Paracelsus (1493–1541) developed the idea that different drugs could be used for different diseases. But he had a strange interest in poisons. When people attacked him for it, he said that poisons are only poisonous if you have a big enough dose. He tested this theory on animals...

Circulation

In the early 1600s, William Harvey showed that blood doesn't just sit in the body, it is pumped round and round by the heart. This was a medical breakthrough, but the way Harvey showed this seems really cruel now. He tied a living dog to a table, then cut it open to show the dog's heart pumping and the blood flowing.

CUTTING AND DRILLING

Today, surgery is done with the latest high tech equipment while you're asleep, but the first surgeons cut into you while you lay screaming in agony. Sometimes, though, they might just have saved your life.

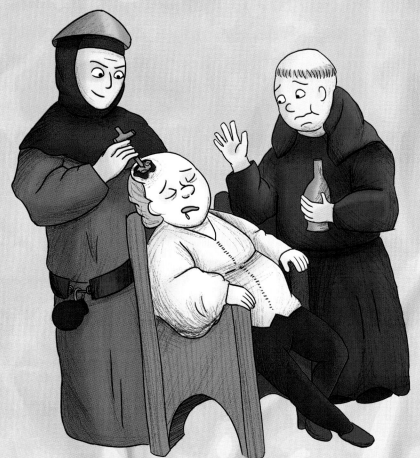

Hole in the head

'Trepanning' dates back at least 8,500 years. It involves drilling a large hole in the skull! It sounds unbelievably painful and very dangerous. Yet, trepanning was practiced right up until a few hundred years ago. No one really knows why. Maybe it was to stop people suffering fits, or to let evil spirits out.

Stitched up!

Surgeons were stitching up serious wounds with a needle and thread at least 6,000 years ago. They'd sew the wound together with a bone needle, and thread made from animal tendons or plants. Now we call this 'suturing'. Back then they probably just said 'aaagh'!

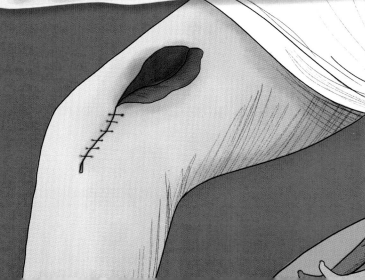

Nose job

In 15th century Italy, if you lost your nose in a sword fight (as many did), you went to Dr Gaspare Tagliacozzi. Tagliacozzi would rebuild your nose for you by **grafting** skin from your arm. This meant going around with your arm sewn to your nose for months. And it wasn't guaranteed to work.

Nose job No. 2

The earliest known plastic surgeon, Sushruta, lived in India 2,800 years ago. Back then, noses were chopped off as a punishment. So Sushruta developed a line in rebuilding noses. He simply cut a flap from your cheek then folded it over to make a new nose, holding it in place with stitches until it was properly grafted (attached).

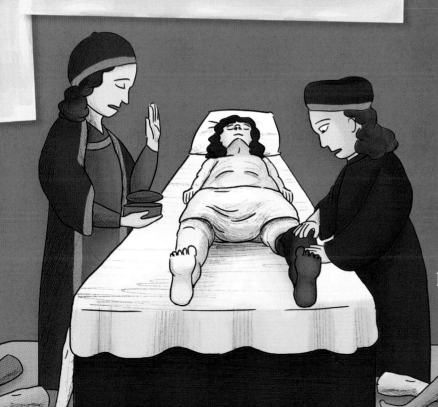

One black leg, one white

Back in the 3rd century, the leg of a church deacon in Constantinople became infected. No problem, said local saints Cosmas and Damian to the deacon, in a dream. They cut off his diseased leg and stitched another in its place, chopped from a newly dead body. But the dead man was black. So when the deacon woke up, he had one black leg and one white! If the story is true, it was the first ever **transplant**...

OLD DOCTORS

In prehistoric times, people relied on magic and traditional knowledge when they fell ill. But most modern doctors try to use scientific knowledge instead. The first specialist doctors appeared in the time of Ancient Egypt, nearly 5,000 years ago.

Stretched

The Greek doctor Hippocrates (see opposite) invented this device, in which the patient was stretched with ropes tied round their arms and legs. It looks pretty nasty, and it inspired the medieval torture device, the rack. However, it was actually designed to help set broken bones properly. Hospitals still use similar 'traction' devices to relieve pressure on damaged backs.

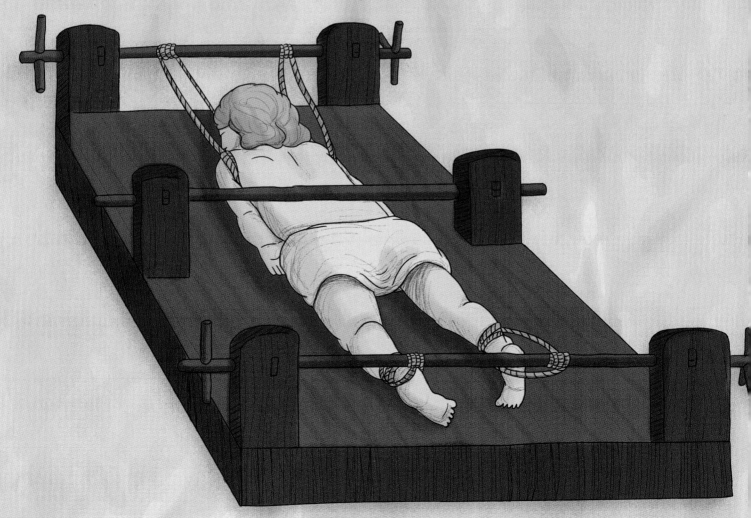

No snakes, please!

Hippocrates lived on the Greek island of Kos, 2,500 years ago. Ancient Greek doctors often killed patients by trying to cure them with snake venom, but Hippocrates knew this was wrong – diseases weren't punishments by the gods but had natural causes. He also said doctors had a duty of care to patients.

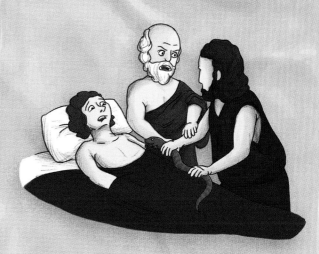

Doctors today still swear a Hippocratic Oath to treat patients well, based on Hippocrates' ideas.

A promising start

The first known doctor was Imhotep, who lived in Egypt 4,600 years ago. Apparently he could diagnose and treat 200 diseases, such as tuberculosis and arthritis. He knew anatomy, too, and maybe even how blood circulates. He was also a brilliant engineer who built pyramids. So if he didn't cure you, he'd give you a good tomb!

Nasty wound, there! Great!

Galen (about 129–200 CE) was the most famous doctor in the Roman Empire. He learned what bodies are really like from the terrible wounds that gladiators suffered in fights. His knowledge became the basis of medicine for the next 1,300 years. Galen boasted: 'I have done as much for medicine as Trajan did for the Roman Empire when he built bridges and roads.'

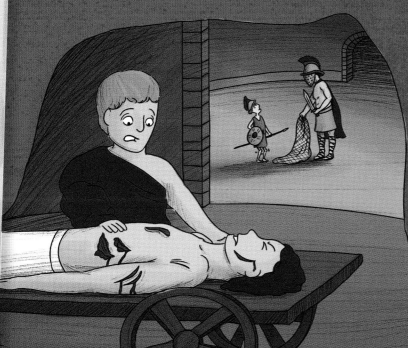

55

LEECHES AND BLOODLETTING

Next time you complain about having to take medicine when you're ill, just think about what you might have had to go through in the past – anything from being covered in bloodsucking slugs to being cut and made to bleed.

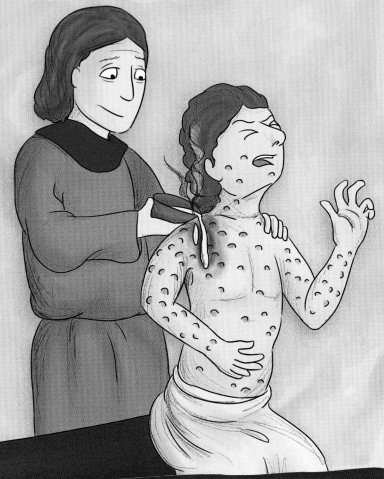

Heavy metal

In the 1490s, the terrible disease syphilis began to spread across Europe. One of its effects was to cover the body in horrible pustules. To treat the disease, doctors spooned the liquid metal mercury on the pustules, or made patients sit in a room filled with mercury fumes. Mercury is poisonous and drove patients mad.

Fancy a cup?

Cupping dates back 5,000 years and is still performed by some people today. It involves heating a cup and pressing it hard onto the skin. The heat creates a suction effect that draws blood to the skin under the cup. The extra blood flow is said to reduce pain. But there is little evidence that it works.

Blowing smoke up

When Europeans brought tobacco back from America in the 1500s, doctors thought it might help treat some ailments. Some doctors had a strange way of using it. In the 1800s they would light a pipe full of tobacco and use a long tube to blow the smoke up the patient's rear end! They called this a 'tobacco **enema**'.

You want blood?

In ancient times, many doctors believed having too much blood in your body made you ill. So whenever people got sick, the doctors would cut open a large blood vessel in the patient's arm or neck to let out blood. Many died or became worse through loss of blood. But the practice went on until the mid-1800s.

Bloodsuckers

You didn't have to cut people to let out blood. Some doctors used leeches to suck it out instead. The great thing about leeches, they thought, is that they can be stuck on close to the diseased organ! Some doctors have recently suggested that leeches might be a good way of treating some ailments after all.

57

BARBERS AND QUACKS

In the past, there were all kinds of words used to describe doctors – not all of them nice! If you had a bad injury and maybe needed a leg cut off, you'd go to the barber surgeon. If you needed drugs, you could try an **apothecary**...

Quack pot

The 18th century was the heyday of 'quacks': people who tried to sell you their own brand of medicine – guaranteed to cure your problem! One notorious quack was American Dr Elisha Perkins (1741–99). Perkins claimed he could cure rheumatism and pain by waving two metal rods or 'tractors' over you.

My legs grew back!

Some quacks made absurd promises. A 19th century cartoon makes fun of the promise of miracle cures made by Morrison's vegetable pills. A man with two wooden legs claims his real legs have grown back, because he's taken Morrison's. The other man is not convinced!

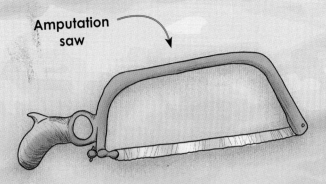

Amputation saw

Sawbones

There was no way to treat a badly injured arm or leg. So the only way to stop the wound going bad and killing you was to remove the wounded part with a saw. The men who did this were called 'sawbones'. Having your leg sawn off was unimaginably painful, since there were no **anaesthetics** back then.

Barber surgeons

In the Middle Ages, you didn't go to a doctor to have an injured arm or leg amputated – you went to the barber! Barbers were good with knives, so they could cut hair or limbs. They could also cut your arm to let blood. That's why their symbol became a red-and-white pole, representing blood running down an arm.

NEW BLOOD

If you lose lots of blood, from an injury or during surgery, your life may be saved by an input of someone else's blood. This is called a 'transfusion'. We now take transfusions for granted, but in the past they could kill you.

Dog to dog

The first ever transfusion was performed on two dogs in 1665 by English physician Richard Lower. He connected an artery in one dog with a vein in the other, via a glass pipe. When he sliced an artery in the second dog so that it lost a lot of blood, it was kept alive by blood flowing from the first dog.

Bad blood

No one understood at the time that not all blood is the same. So, in 1667, a doctor in Paris, Jean-Baptiste Denys, tried to give a man sheep's blood in the same way Lower had transfused two dogs. But the sheep's blood killed the man. Denys was tried for murder and transfusions were banned.

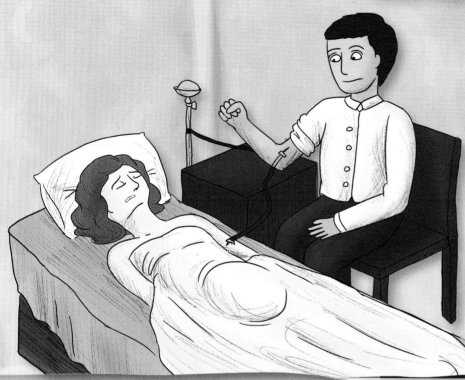

The first human transfusion

In the early 1800s, Dr James Blundell was appalled by how many women died from loss of blood during childbirth.

So in 1818, he used a syringe to inject a mother, who had lost a lot of blood, with blood collected from her husband's arteries. The transfusion worked and the mother survived.

Blood types

Blundell's success was a lucky one-off, and most transfusions performed afterwards killed the patient. Then in 1900 Austrian-American doctor Karl Landsteiner (1868–1943) discovered why. Blood belongs to three different groups – A, B and O – and for a transfusion to work, the blood must be of the same kind.

Blood bank

The first transfusions were made by hooking the donor up directly to the patient. But then it was discovered that blood could be stored for several days in a refrigerator. In World War I, stores of blood called blood banks were set up, saving the lives of the countless wounded soldiers.

61

SICK HOSPITALS

The first proper hospitals appeared in India some 2,400 years ago, and cared for the sick very well. But sometimes in the past, hospitals were horrible places that were as likely to kill you as cure you.

Holy hospitals!

Some of the first hospitals in Europe were in nunneries. They could be terrible places, where you might pray to get out alive. In the infamous Hôtel-Dieu in Paris in the 18th century, several people would be crammed into each bed, and patients with infectious diseases mixed with the mentally ill. Very few survived.

Lady with the lamp

Nurse Florence Nightingale was shocked by the crowded, dirty conditions in field hospitals for British soldiers, wounded in the Crimean War, in the 1850s. Her insistence on sanitary conditions, and taking close personal care of patients – even at night – helped change hospitals from places where people went to die to places of healing.

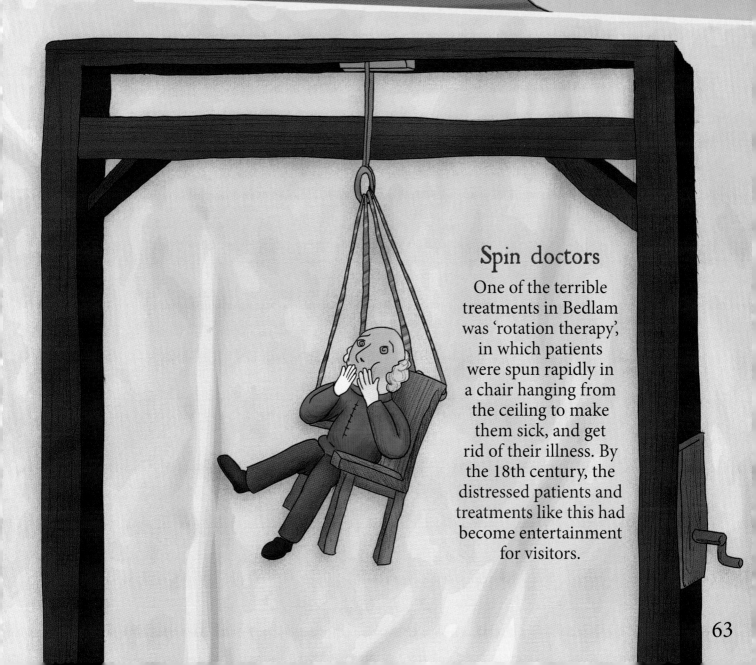

Bedlam!

Set up in 1247, the Bethlehem hospital in London was the first hospital for the mentally ill in Europe. It was once a house of horrors, where distressed patients were chained up in appalling conditions. Their wailing was so horrific that the hospital's nickname, 'Bedlam', came to mean dreadful noise and chaos.

Spin doctors

One of the terrible treatments in Bedlam was 'rotation therapy', in which patients were spun rapidly in a chair hanging from the ceiling to make them sick, and get rid of their illness. By the 18th century, the distressed patients and treatments like this had become entertainment for visitors.

63

CAN'T FEEL A THING

Without anaesthetics, major operations, such as heart surgery, would be impossible. Anaesthetics are chemicals that send you to sleep (general anaesthetic) or dull the pain in the affected area (local anaesthetic), while surgery is performed.

Sleepover

In 1847, Scottish doctor James Simpson had a party and tried out the anaesthetic, chloroform, on two of his doctor friends. It knocked all three of them out. Soon chloroform was being widely used as an anesthetic during operations. It was later found that chloroform is slightly poisonous.

ZZZZZZZZZ

It's a knockout

In Massachusetts in 1846, American dentist William Morton pulled a tooth out while his patient was fast asleep. This was the first ever successful operation under general anaesthetic. The anaesthetic he used was a chemical called 'ether', which was warmed in a jar to create fumes that the patient breathed in.

Ether dispenser

Ether fumes

Sponge soaked in ether

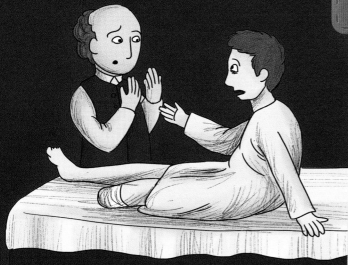

What a laugh!

In the 1830s, people went to demonstrations of the effects of nitrous oxide, known as 'laughing gas' because it makes you laugh and feel pain less. In 1844, dentist Horace Wells made patients breathe laughing gas from a pig's bladder and then painlessly pulled teeth out. However, other dentists found patients cried out in pain halfway through.

When are you going to begin?

Just a few months after Morton's tooth extraction, Scottish surgeon Robert Liston amputated a patient's leg while he was entirely unconscious from ether. The patient came round, entirely unaware the operation had been performed, asking Liston, 'When are you going to begin?'

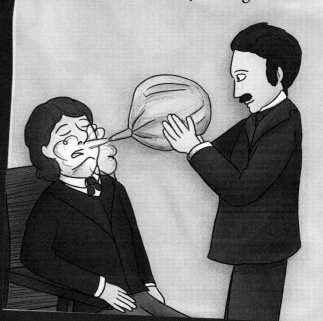

Dangerous darts

Some native South American people blow poison darts dipped in a plant extract called 'curare'. Curare paralyses victims so completely that they stop breathing and die. In the past, curare was tried as a general anaesthetic, until it was discovered that although it made patients immobile, they could still feel the pain of an operation!

67

ANCIENT MEDICINE

Many herbs and natural materials have healing properties, and people learned long ago that these natural medicines can often be made more effective by preparing them or mixing them in certain ways. This ancient art of pharmacy dates back to Ancient Egypt and beyond.

The father of pharmacy

The Ancient Egyptians believed pharmacy began with the god Horus. Horus lost his eye in an epic battle, but it was healed by the god Thoth. Pharmacists today write the sign 'Rx' on prescriptions for medicine. Some believe they are writing the ancient sign for the eye of Horus. Others think that Rx is just medieval shorthand for the Latin for 'recipe'.

Herb master

For 1,500 years *De Materia Medica* ('About Medical Materials') was the 'bible' on medical herbs. This vast book was written by the Greek pharmacist Dioscorides in the 1st century CE. It describes more than 1,000 herbal medicines, many discovered by Dioscorides himself as he followed Roman armies round Europe.

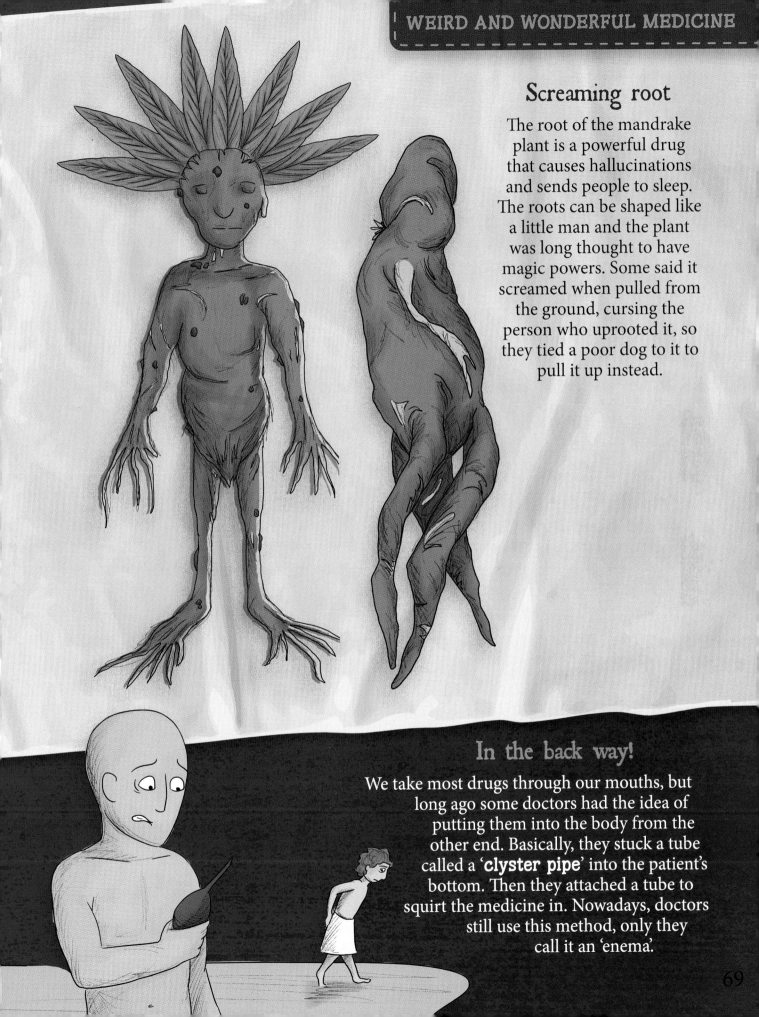

Screaming root

The root of the mandrake plant is a powerful drug that causes hallucinations and sends people to sleep. The roots can be shaped like a little man and the plant was long thought to have magic powers. Some said it screamed when pulled from the ground, cursing the person who uprooted it, so they tied a poor dog to it to pull it up instead.'

In the back way!

We take most drugs through our mouths, but long ago some doctors had the idea of putting them into the body from the other end. Basically, they stuck a tube called a '**clyster pipe**' into the patient's bottom. Then they attached a tube to squirt the medicine in. Nowadays, doctors still use this method, only they call it an 'enema'.

69

CHINESE ROOTS

The Chinese have an ancient tradition of looking for cures in nature, but they don't just use herbs for medicine. They have used all kinds of other strange things, too, from scorpion stings to centipedes.

Live forever!

For many centuries, Chinese chemists searched for a pill that would make people live forever. One story tells that a man called Wei Bo-yang did succeed in making an '**immortality**' pill. The story also says that the legendary Yellow Emperor found this pill and so lived on forever.

The point of medicine

Acupuncture involves sticking sharp needles into the skin. Ouch! The needles are meant to activate healing pathways or 'meridians' in the body. Acupuncture has been practiced in China for 2,000 years, and it may have been known in Europe long ago, too. 'Ötzi', a man from 5,000 years ago, found frozen in mountain ice, has tattoos that seem to mark acupuncture meridians.

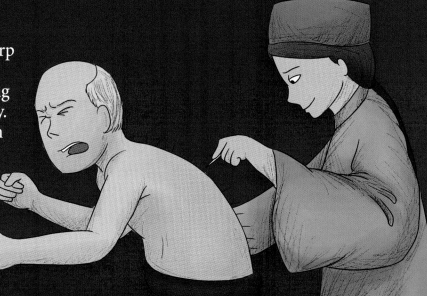

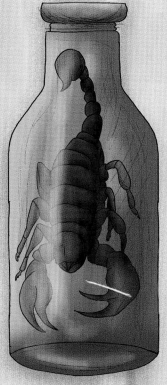

This medicine stings!

Most of us would steer well clear of scorpions. The sting in their tail is always painful, and can be deadly. Yet the Chinese have long believed that pickled scorpions, or quan xie, can cure fits, headaches and swelling. But maybe patients just said 'I'm better!' as soon as they saw the medicine they were being given!

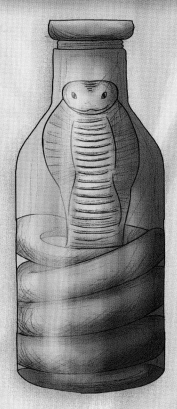

Snake wine

If you're feeling a little tired, or you're losing your hair, what you need is snake wine, according to ancient Chinese medicine. Snake wine is made by drowning a venomous snake in rice wine! It was first used 2,700 years ago.

Killing seahorses

Seahorses are beautiful marine creatures that are in danger of extinction. Yet those who believe in traditional Chinese medicine buy 25 million of them every year. They are convinced dried seahorses can cure kidney complaints, wheezing, tummy aches and much more. There is no scientific evidence for this.

Dragon bones

The fossilized bones of long-dead creatures are often called dragon bones in China. Ground to a powder, they are used in Chinese medicine to cure heart problems, stress and fever. In the ancient past, dragon bones, or 'oracle bones', were also scratched with questions that people wanted to ask the gods.

71

IT'S CHEMISTRY

Most modern medicines are made from chemicals rather than herbs, and we owe this to the scientists of Ancient Islam. It all started with the brilliant Jabir ibn Hayyan, also known as Geber, who may have lived in Kufa (in modern Iraq) 1,300 years ago.

Bedside manner

To get just the right mix of chemicals to make the medicine, the Islamic physicians believed they needed to spend time with each patient and find out all they could about their symptoms and character.

Take your poison

Jabir knew all there was to know at that time about the chemistry of poisons. In his famous Book of Poisons, he described hundreds of toxic substances, and how they react in the body. He also described '**antidotes**' – substances that work against the effects of poison – and gave practical demonstrations of them in action.

Going to the chemist

Ancient Islamic cities such as Baghdad were packed with pharmacies, all stocked with jars of various mixtures of chemicals. Today, we are used to treating illnesses with drugs – that is, substances that affect the chemistry of the body. Back then it was exciting and new.

Poison, madam?

Many medicines included deadly poisons, such as henbane, hemlock and black nightshade. Ancient chemists weren't trying to kill the sick. In fact, they'd made a crucial discovery: what matters with giving medicines is the dose. In small doses, some poisons are powerful drugs for relieving pain or sending you to sleep. However, they didn't always get the dose right. Bye bye!

73

GREEN MEDICINE

The chemicals and exotic ingredients used by apothecaries were hard to come by or very expensive. So for centuries, most people relied on herbs for their medicine. These could be picked from the wild or grown in 'physic' gardens.

Hildegard's herbs

One of the most famous books on herbal medicine was written by a 12th-century German nun, Hildegard of Bingen, who also happened to be one of the first known musical composers. Hildegard believed in 'green power', using plants and herbs to cure a number of illnesses. Her books are still referred to today.

Garden of medicine

Monasteries set up physic gardens for growing herbs to treat the sick, as well as provide food seasonings and dyes. There were different areas for plants to treat coughs and colds, liver and bladder complaints, digestion, headaches, anxiety and depression, and other conditions.

Powerful burdock

Nicholas Culpeper wrote one of the best books on herbal medicine in 1653, but even Culpeper had some odd ideas. He thought that the herb burdock would prevent farting and rabies, and also cure snake bites. He even thought that if women wore it on their heads it would stop their wombs collapsing…

Look-alikes

It was hard to know what substances would heal what ailments. Many people believed in the 'doctrine of signatures'. This was the idea that plants and foods would work as treatments for parts of the body that they look like. So since a sliced mushroom looks like an ear, it must be good for hearing.

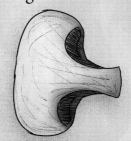

Slice of tomato – for the heart

Slice of carrot – for eyes

Slice of mushroom – for hearing

Whole carrot – for the nose

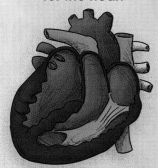
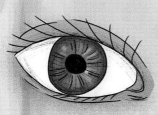
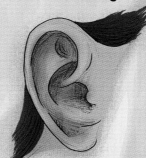
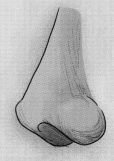

Wise woman

In many country villages, if someone was ill, they went to see the witch, who knew all about herbs. For some people, a witch was just a wise old woman who knew how to heal the sick, but other people were afraid of witches' skills and believed they served the devil.

WILLOW POWER

If someone has a bad headache or another pain, it's easy for them to take an aspirin or other painkiller. But it wasn't always that way...

Barking mad

Aspirin is made from the chemical **salicin**. A form of salicin occurs naturally in the bark of willow trees, myrtle and meadowsweet plants. But getting it out of a tree is a real headache...

Taming the snake

Ague (probably malaria) was so nasty that people pictured it as a snake gripping you! Chewing on Peruvian bark helped, but this wasn't easy to get in England. Then in 1758, Oxfordshire vicar Edward Stone noticed that willow bark tasted a bit like Peruvian bark. He tried it on a few ague victims and they felt better at once.

Drugged up

In the 1800s, if someone was in pain, doctors would prescribe a glass of **laudanum**. This certainly eased the pain, but it was a strong, addictive drug, related to heroin. Many who took it became desperate drug addicts, including many women who took it simply to ease period pains.

Aspirin

The pain-relieving ingredient in willow bark is a chemical called salicin. Salicin is acidic and attacks the stomach. In the 1890s, Bayer company scientists made the wonder drug aspirin by modifying salicin to make it less damaging. No pain, big gain.

Aspirin

FINDING A MAGIC BULLET

Syphilis is a terrible disease that once ravaged Europe. There seemed to be no cure – until a German physician found a chemical that killed the germs but left the patient unharmed. For him, it was like the mythical 'magic bullet' that kills only the baddie.

We've won!

French king, Charles VIII, was rather pleased when he marched his conquering army into Naples, Italy in 1495. But his soldiers caught the horrific disease syphilis in the city and spread it across Europe.

Nose job

Syphilis didn't often kill people straight away, but its effects were really horrible. It made the body burst out with boils so foul that syphilis came to be known as the Great Pox. It also rotted away the nose, so that many syphilis victims had to have artificial ones. It could make the sufferers mad, too.

Fume it out

People were so desperate to be cured of syphilis that they tried anything. The metal mercury was thought to help, and one idea was to sit in a box filled with mercury fumes. Mercury is poisonous, and its effects were painful, unpleasant and even fatal.

The deaf composer

The brilliant composer Ludwig van Beethoven (1770–1827) famously went deaf as he got older, so it became very hard for him to write music. People have wondered if his deafness was caused by syphilis. Other syphilis victims from history include Henry VIII, Adolf Hitler and the Russian writer Leo Tolstoy.

On target

The German scientist Paul Ehrlich (1854–1915) believed that if bacteria could be stained by certain dyes, then they might also be killed by them, as if by magic bullets. He and his colleagues tried hundreds of stains and in 1914 they found one that worked against syphilis germs. They used it to make the drug Salvarsan, the first effective treatment for the disease.

SUGAR IN THE BLOOD

A person who has diabetes can't keep down the sugar levels in their blood. The problem lies with the body chemical **insulin**, which should control sugar levels, but doesn't. That's why diabetics (people with diabetes) must receive regular injections of insulin.

Sweet pees

In the Middle Ages, many doctors knew just how to diagnose diabetes. They tasted the patient's pee. If it tasted sweet, the patient had diabetes. But they hadn't much idea how to treat the disease. Some suggested drinking wine, some eating a lot of sour food, and others riding horses!

Finding insulin

In 1889, German scientists Oscar Minkowski and Joseph von Mering removed a dog's pancreas, for their studies of digestion. Later they saw flies feeding on the dog's pee, found it was sweet and discovered that it is the pancreas that controls sugar levels. In 1921, Canadian Frederick Banting and American Charles H. Best discovered that it does this by making insulin.

Bacteria factory

In the 1970s, scientists found how to use bacteria as factories for making human insulin. They inserted the gene for human insulin into E. coli bacteria, then put them in a huge vat to multiply. As they multiplied, the bacteria followed their new genes' instructions and made lots of human insulin. This is now the main source of insulin for diabetics.

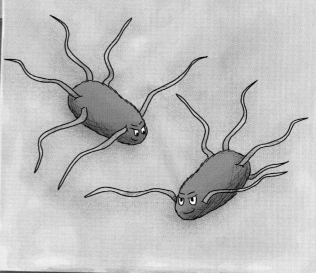

Sweet pigs

The discovery of insulin was a major breakthrough for the treatment of diabetes. Insulin could be extracted from the pancreases of pigs and cows killed for meat. Then it was refined and given to diabetics in injections that kept their condition under control.

Out of the box

Pandora was the mythical girl who was so curious that she opened a box and let out things that would hurt mankind, such as war and disease. Some scientists were worried that genetically modified (GM) bacteria for making insulin might escape and spread – just like 'opening Pandora's box'. At a conference in 1975, they agreed they should never work with disease-causing bacteria, and always work in a completely secure room.

83

Since Fleming discovered penicillin, scientists have discovered 150 other 'antibiotics' – drugs that attack bacteria. Most antibiotics are now entirely man-made but, as with pencillin, many were often originally found in natural sources. The antibiotics, tetracycline and streptomycin, both came from bacteria, and the natural world is full of antibiotic substances that may inspire new drugs. Here are some of the weirdest places that scientists are looking...

Alligator blood

Scientists were baffled that alligators seem to get badly wounded in fights with other alligators, yet their wounds didn't become infected. When they investigated, they found that alligator blood contains natural antibiotics that are effective against a wide range of infections.

Catfish slime

Catfish seem to survive injuries without getting infected, and scientists recently found out why. Their bodies are covered in a kind of slime, rich in antibiotics that seems to be good at killing off germs, such as Klebsiella pneumoniae – which attacks the lungs – and E. coli.

Frog skin

Frogs can survive in water that would kill other creatures. Scientists now know one reason why: their skins are covered in 100 bacteria-killing substances. Yet most are also dangerous to humans, so scientists must find a way to apply their germ-killing powers without hurting us!

Panda blood

One of the most powerful of all antibiotics, cathelicidin-AM, occurs in panda blood. It kills off germs in a fraction of the time it takes most other antibiotics. Pandas are very rare, so scientists make synthetic panda blood to conduct experiments with cathelicidin-AM.

Leafcutter ants

Leafcutter ants are known for their supersize strength in carrying leaves, but they also have superpowers against germs. Scientists have found that their bodies deploy multiple chemicals to fight bacteria and fungi, just as doctors use multi-drug approaches to treat difficult infections.

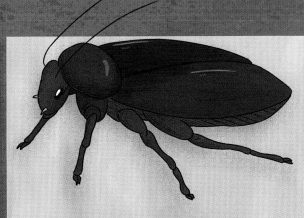

Cockroach brains

Crushed cockroach brains contain nine different kinds of antibiotic. Scientists are trying to find out if some may be used to treat E. coli infections or even MRSA (infection by the S. aureus 'superbug' bacterium that has become resistant to other antibiotics).

GRISLY FACTS: GRUESOME SYMPTOMS

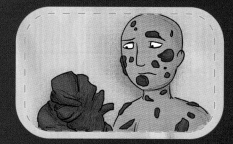

Epidermodysplasia verruciformis is a very rare disease that makes giant warts grow thickly all over the body.

When you cough, germs can travel about ten feet (3 m) if you don't cover your nose or mouth with your hand or a handkerchief.

Buboes are horrible pus-filled blisters that erupt under the arms or on the neck or groin. They are a sign of bubonic plague.

Bacteria in your nose and mouth are what give you bad breath.

If a person were bitten by a dog with rabies, they might start dribbling and foaming at the mouth with saliva.

Leprosy has horrific effects, such as making bits of the body drop off. People suffering from the disease were once cut off from other people and forced to live in 'leper colonies'.

When someone throws up it often seems to contain carrots, even though they haven't eaten any! These are actually parts of the stomach lining that have come off.

Some soldiers in the American Civil War had 'glow-in-the-dark' wounds because of a bioluminescent bacteria that was puked up by nematode worms.

A single sneeze can spray out 6 million little viruses into the world.

For headaches and low spirits, medieval herbalists gave concoctions including the herbs betony and vervain. Scientists have now discovered that both of these contain chemicals that are good for treating migraines and depression.

According to one ancient papyrus, the Ancient Egyptians used donkey, dog, gazelle and fly poo as medicines, taken through the mouth.

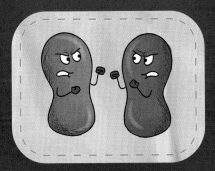

Since the discovery of antibiotics, more and more bacteria have developed resistance to them, primarily through overuse.

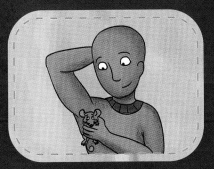

The Ancient Egyptians used lizard blood, dead mice, mud and mouldy bread as ointments for sores.

According to legend, the Ancient Greek king Mithridates developed an antidote to all poisons. The Romans turned it into 'theriac', a drug that would cure all ailments. It didn't work, but people continued using it until the late 1800s.

Recently, scientists tested a medieval eye cure made from onions, garlic, wine and bull's gall bladder juice. It looked yucky but had amazing antibacterial properties.

GRISLY FACTS: DOCTORS MAKE ME SICK!

The Ancient Egyptians thought they could cure toothache by slitting open the belly of a mouse and laying its still-warm body on your gums. Uggh!

In the past, doctors stopped severe bleeding, perhaps after an amputation, by cauterizing the wound – scorching it with red-hot metal or boiling tar. Aaagh!!!

In 1738, the British parliament paid quack Joanna Stephens £3,000 for her bladder stone cure, and the Prime Minister Robert Walpole ate 175 lbs (80 kg) of it – but it was just soap and eggshells. Frothy!

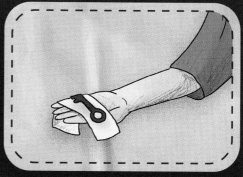

If you were bitten by a dog with rabies, a priest might scorch the bite with a red-hot key or nail, called St Hubert's Key. Surprisingly it could actually work if done soon after the bite, by killing the virus. Ouch!

In the Middle Ages, if a doctor offered you a clyster to treat your illness, run! A clyster was a tube that he stuck up your bottom and then poured in hot water, or pig's bile. Eeeegh!

To cure a sore throat in the Middle Ages, a doctor might tell you to eat dog poo. No need to eat it wet – dried and mixed with honey was fine... Mmm!

One species of bacterium is so resistant to radiation that scientists have nicknamed it 'Conan the Bacterium'.

When you flush a toilet, an invisible cloud of water full of germs shoots far up into the air.

Poo, by weight, is mostly bacteria.

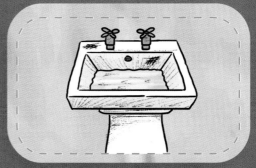

The average kitchen sink contains 100,000 times more germs than the bathroom.

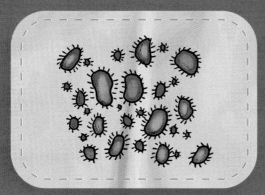

Bacteria double in number every 20 minutes. A single bacterium could divide and multiply into trillions in just one day, if all the bacteria survived.

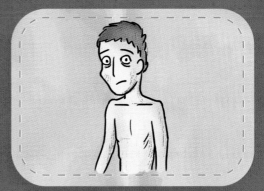

There are more bacteria on your skin than there are people in the world. Take out the water, and a tenth of your body weight is bacteria.

GLOSSARY

Acupuncture	An ancient Chinese treatment involving inserting needles into the body.
Amber	The resin that oozed from trees long ago and has turned solid.
Anaesthetic	A chemical that numbs pain or sends you to sleep.
Antibiotic	A drug that attacks bacteria.
Antidote	A remedy for counteracting the effect of a poison or germ.
Antiseptic	A substance wiped or sprayed on surfaces to kill germs.
Apothecary	Someone who mixed and sold medicines.
Ayurveda	An ancient medical system, developed in India, that treats people with complex mixes of herbs.
Bacteria	Microscopic living things, made from just a single living cell. A small number of them cause disease (singular: bacterium).
Bile	A dark green fluid made by the liver, and a passionate, impulsive humor.
Bloodletting	Cutting the skin to let out blood in the vain hope of curing illness.
Bubo	A huge boil-like eruption on the skin, caused by bubonic plague.
Clyster pipe	A pipe for injecting fluids and gases up a patient's bottom for medical purposes.
Consumption	The old name for the killer lung disease tuberculosis (TB).
Cupping	Applying heated cups to the skin to draw the blood to the surface.
Diabetes	A disease caused by fluctuations in sugar levels in the blood.
Doctrine of signatures	The idea that treatments for different parts of the body can be identified by similar shapes in nature.
Enema	The injection of fluid or gas into the bottom for medical purposes.
Epidemic	A widespread outbreak of an infectious disease.
Germ	A microbe that causes disease when it enters the body.
Grafting	Attaching tissue from someone else, or from another part of your own body.
Humor	One of four states of the body that old medical theory believed needed to be in

balance to maintain good health.

Hygiene	Cleanliness that reduces the spread of infection.
Immortality	Living forever.
Immune system	The body's own cells that make it able to protect iteself against infection.
Inoculation	Using germs to stimulate the immune system to guard against infection.
Insulin	The body chemical that regulates levels of sugar in the blood.
Laudanum	An addictive painkilling drink made by mixing the drug opium in alcohol.
Miasma	A damp, smelly mist that was believed to spread disease.
Microbe	A microscopic living thing, especially one that causes diseases, such as a bacterium.
Mosquito	A tiny blood-sucking insect that can spread diseases, such as malaria and yellow fever.
Pandemic	A huge worldwide outbreak of an infectious disease.
Penicillin	A substance in the Penicillium mould that kills bacteria.
Phlegm	The slimy stuff produced when you have a cold, and a shy, reserved humor.
Plasmodium	A tiny organism that causes the disease malaria.
Quinine	A substance found in the bark of the cinchona tree, used for treating malaria.
Salicin	A substance found in meadowsweet and the bark of willow trees, used to reduce inflammation, and also to make aspirin.
Transfusion	Transferring blood from one person to another.
Transplant	Replacing one person's faulty body organs with another's healthy organs.
Vaccination	Using dead or inactivated germs to stimulate your body's immune system, to guard against infection.
Virus	A tiny germ that reproduces only inside other living cells – including yours.

INDEX

The Author

John Farndon is the author of many books on science, technology and nature, including the international best-sellers Do Not Open and Do You Think You're Clever? He has been shortlisted five times for the Royal Society's Young People's Book Prize for a science book.

The Illustrator

Venitia Dean grew up in Brighton, UK. She has loved drawing ever since she could hold a pencil. After receiving a digital drawing tablet for her 19th birthday she transferred to working digitally. She hasn't looked back since!